ODD BUT ADORABLE ANIMALS

Jerboas

Elsie Olson

Big Buddy Books

An Imprint of Abdo Publishing
abdobooks.com

abdobooks.com

Published by Abdo Publishing, a division of ABDO, PO Box 398166, Minneapolis, Minnesota 55439.

Printed in the United States of America, North Mankato, Minnesota
102024
012025

THIS BOOK CONTAINS RECYCLED MATERIALS

Design: Elena Klinkner, Mighty Media, Inc.
Production: Mighty Media, Inc.
Editor: Liz Salzmann
Cover Photograph: Lauren Suryanata/Shutterstock Images
Interior Photographs: blickwinkel/Alamy Stock Photo, p. 19 (top); ChrWeiss/Adobe Stock, p. 21; edmon/Shutterstock Images, p. 25; Iman.Ebrahimi/Shutterstock Images, p. 11; Lauren/Adobe Stock, p. 9; Lauren Suryanata/Shutterstock Images, pp. 5, 18, 27; mgkuijpers/Adobe Stock, p. 19 (bottom); Purchase, Edward S. Harkness Gift, 1926/Metropolitan Museum of Art, p. 23; stas111/Adobe Stock, p. 14 (compass rose); tanarch/Adobe Stock, pp. 14–15 (maps); Yerbolat Shadrakhov/Shutterstock Images, pp. 7, 13, 17, 29
Design Elements: flovie/Shutterstock Images (patchwork pattern); Mighty Media, Inc. (series logos & icons)

Library of Congress Control Number: 2024938321

Publisher's Cataloging-in-Publication Data
Names: Olson, Elsie, author.
Title: Jerboas / by Elsie Olson
Description: Minneapolis, Minnesota : ABDO Publishing, 2025 | Series: Odd but adorable animals | Includes online resources and index.
Identifiers: ISBN 9781098295134 (lib. bdg.) | ISBN 9798384915188 (ebook)
Subjects: LCSH: Jerboas--Juvenile literature. | Rodents--Juvenile literature. | Omnivorous animals--Juvenile literature. | Desert animals--Juvenile literature. | Nocturnal animals—Juvenile literature. | Curiosities and wonders--Juvenile literature.
Classification: DDC 599.35--dc23

Contents

A Jerboa Encounter

You're on a **twilight** hike in the Middle East. Suddenly a big-eared, big-eyed creature jumps out in front of you. It looks like a kangaroo. But it's the size of a mouse. Before you can take a picture, the critter twitches its ears and hops away. You've just met the odd but adorable jerboa!

Jerboa is pronounced jer-BO-eh.

A Closer Look

Jerboas are a type of **rodent**. There are 33 **species** of jerboa. Most are between 2 and 6 inches (5 and 15 cm) long. Jerboas have long tails. The tails are between 3 and 10 inches (7 and 25 cm) long. That's much longer than their bodies!

A jerboa's tail helps the animal balance and stay upright.

A jerboa's big ears help keep it cool. Jerboas can also hear danger from far away. They have short front legs and long back legs. This allows them to jump long distances. Jerboas can leap up to 10 feet (3 m) with one jump! Their strong legs make jerboas expert diggers.

Jerboas leap to escape from predators.

Desert Dwellers

Jerboas are desert creatures. They live in deserts and **grasslands** in eastern Europe, Asia, and northern Africa. These **habitats** are often very hot. Jerboas stay cool in underground **burrows**. In the summer, jerboas block their burrow entrances with sand or soil. This keeps moisture in and hot air out!

Most jerboas have light-brown fur. This helps them blend into their desert habitat.

It gets cold in some places where jerboas live. Mongolian jerboas live in the Gobi Desert. This region has very cold winters. Jerboas **hibernate** to survive the long winters. The **rodents** dig **burrows** up to 10 feet (3 m) deep. Then they block the entrances with sand. This keeps the cold out!

Jerboas have fur on the soles of their feet. The fur helps their feet grip the sand.

Location Station

Jerboa Habitat
KAZAKHSTAN
MONGOLIA
UZBEKISTAN
KYRGYZSTAN
TURKMENISTAN
TAJIKISTAN
RBAIJAN
CHINA
AFGHANISTAN
IRAN
PAKISTAN
UWAIT
QATAR
UNITED ARAB EMIRATES
OMAN
YEMEN
INDIAN OCEAN
SOMALIA
ARCTIC OCEAN
NORTH AMERICA
United States
EUROPE
ASIA
ATLANTIC OCEAN
PACIFIC OCEAN
PACIFIC OCEAN
AFRICA
SOUTH AMERICA
INDIAN OCEAN
ATLANTIC OCEAN
AUSTRALIA
SOUTHERN OCEAN

Twilight Treats

Jerboas are most active at **twilight**. During the day, they rest in their **burrows**. In the evening, jerboas search for food. They eat seeds, plants, and insects. Jerboas must always be aware of predators. Owls, cats, foxes, and snakes eat jerboas. Jerboas leap in zig-zag patterns to confuse predators. This gives them a better chance of escaping.

The jerboa uses its hoglike nose to sniff out and dig for food.

Creature Feature

The long-eared jerboa has the largest ears compared to its body size of any animal in the world.

Jerboas don't drink water. They get all the **hydration** they need from their food.

A jerboa can hop up to 15.5 miles per hour (25 kmh).

Jerboas have three to five toes on each foot, depending on the **species**.

Precious Pups

Jerboas spend most of their lives alone. But they have babies twice a year. Female jerboas have two to six pups at a time. A jerboa mother cares for her pups until they can find food on their own. By 11 weeks, jerboa pups can leap. At 14 weeks, they are full-grown.

Jerboas live up to six years.

Critter Culture

Jerboas have been important to human **culture** for thousands of years. Ancient Egyptians created sculptures of the tiny desert creature. Desert communities **admired** the jerboa for its ability to survive its harsh **environment**.

Jerboa figurines have been found in ancient Egyptian temples.

Threats and Hope

Jerboas are secretive animals. Scientists aren't even sure how many jerboas are in the wild! But most **species** are not **endangered**. Still, jerboas face risks. One is **habitat** loss. Human communities are **expanding** into places where jerboas live. But there are also people working to protect jerboa habitats.

Climate changes, such as droughts and heat waves, also affect a jerboa's ability to survive.

Perfect Pets?

Some people capture jerboas or raise them to sell as pets. But jerboas are hard to care for. And they need a lot of space to hop around. It is illegal to keep jerboas as pets in the United States. This is because jerboas can carry **diseases** that are harmful to humans.

Many people want jerboas as pets because they are cute. But jerboas are too shy to make good pets.

Odd or Adorable?

Jerboas are unusual and beloved animals. They look strange, but they're also considered cute by many people around the world. What do you think makes an animal odd? What makes an animal adorable? Do you think jerboas are odd, adorable, or both? Why?

Jerboas have large
eyes that help them
see well in the dark.

Glossary

admire—to regard with pleasure, wonder, and approval.

burrow—an animal's underground home.

culture (KUHL-chuhr)—the arts, beliefs, and ways of life of a group of people.

disease—a sickness.

endangered—having few left in the world.

environment—all the surroundings that affect the growth and well-being of a living thing.

expand—to get bigger.

grassland—a large area of grass with little or no trees.

habitat—a place where a living thing is naturally found.

hibernate—to sleep or rest during the winter months.

hydration—being supplied with enough fluid or water.

rodent—any of several related animals that have large front teeth for gnawing. Common rodents include mice, squirrels, and beavers.

species (SPEE-sheez)—living things that are very much alike.

twilight—early morning or late evening.

Online Resources

Booklinks NONFICTION NETWORK FREE! ONLINE NONFICTION RESOURCES

To learn more about jerboas, please visit **abdobooklinks.com** or scan this QR code. These links are routinely monitored and updated to provide the most current information available.

Index